I0813142

SPACE OBJECTS

PLANETS

by Elizabeth Andrews

Cody Koala
An Imprint of Pop!
popbooksonline.com

Hello! My name is Cody Koala

This book is filled with videos, puzzles, games, and more! Scan the QR codes* while you read, or visit the website below to make this book pop.

popbooksonline.com/obj-planet

*Scanning QR codes requires a web-enabled smart device with a QR code reader app and a camera.

abdobooks.com
Published by Pop!, a division of ABDO, PO Box 398166, Minneapolis, Minnesota 55439.

Printed in the United States of America, North Mankato, Minnesota.
102024
012025

Cover Photo: Shutterstock Images
Interior Photos: NASA, Shutterstock Images, Wikimedia Commons
Editor: Grace Hansen
Series Designer: Victoria Bates

Library of Congress Control Number: 2024938611

Publisher's Cataloging-in-Publication Data
Names: Andrews, Elizabeth, author.
Title: Planets / by Elizabeth Andrews
Description: Minneapolis, Minnesota : Pop!, 2025 | Series: Space objects | Includes online resources and index
Identifiers: ISBN 9781098246983 (lib. bdg.) | ISBN 9781098247546 (ebook)
Subjects: LCSH: Outer space--Exploration--Juvenile literature. | Planets--Juvenile literature. | Solar system--Juvenile literature. | Astronomy--Juvenile literature. | Universe--Juvenile literature.
Classification: DDC 523.2--dc23

Table of Contents

Chapter 1

The Solar System

Our **solar system** was formed around 4.6 billion years ago. **Gravity** pulled together gas and dust to create the Sun. Space objects such as planets formed from leftover gas and dust.

Our solar system is made up of the Sun, eight planets, and thousands of other smaller objects.

Watch a video here!

Chapter 2

What Are Planets?

Planets formed after the Sun. Gas and dust **orbiting** the Sun crashed into each other and stuck together. As they got bigger, their **gravity** pulled the materials into a **spherical** shape.

Learn more here!

A planet is defined by three things. First, a planet must orbit a star. Second, its gravity must be strong enough to make it round.

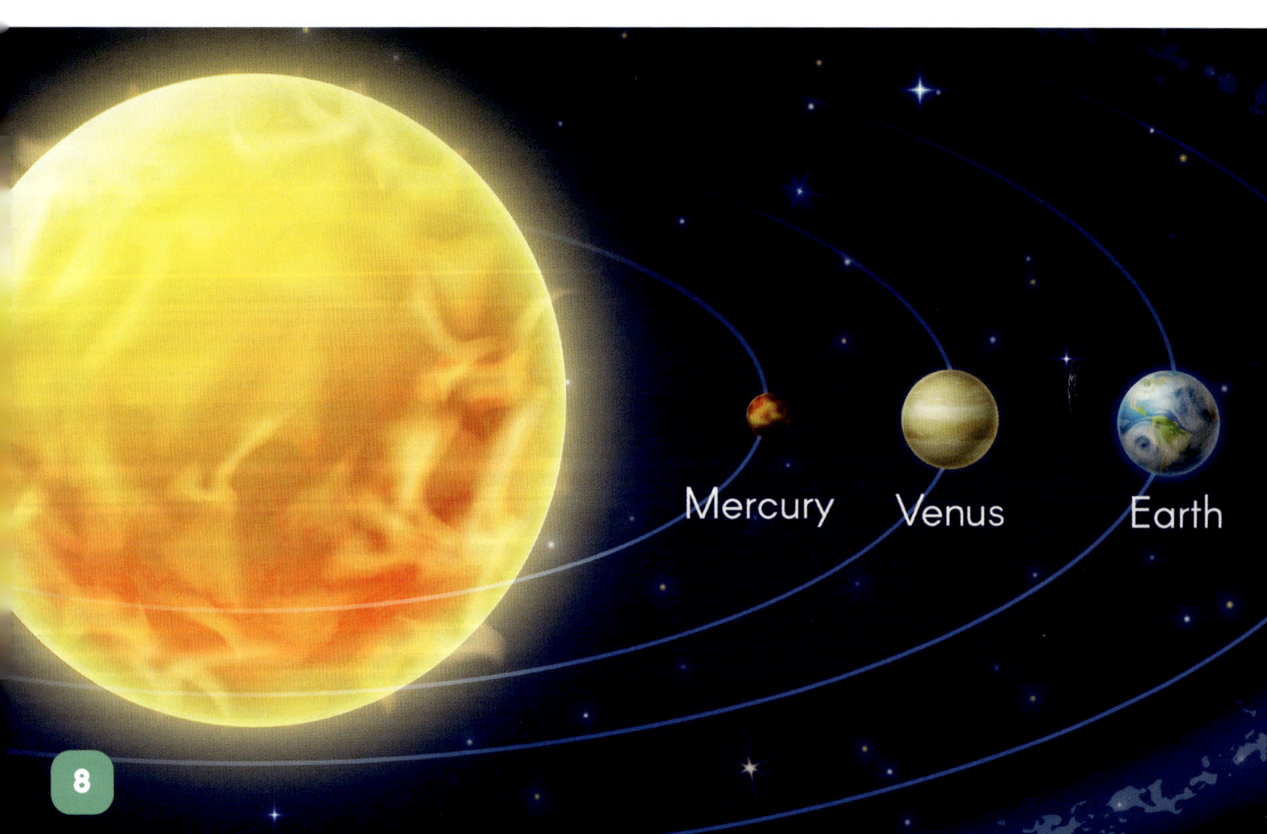

Lastly, it cannot share an orbit with another planet. Eight planets orbit the Sun in our **solar system**.

LENGTH OF DAYS & YEARS ON EACH PLANET

Mercury

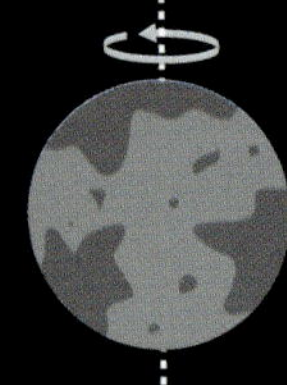

Year:
88 Earth Days
Day: 1,408 Hours

Venus

Year:
225 Earth Days
Day: 5,832 Hours

Earth

Year:
365 Days
Day: 24 Hours

Mars

Year:
687 Earth Days
Day: 24.6 Hours

Jupiter

Year:
4,333 Earth Days
Day: 10 Hours

Saturn

Year:
10,759 Earth Days
Day: 10.5 Hours

Uranus

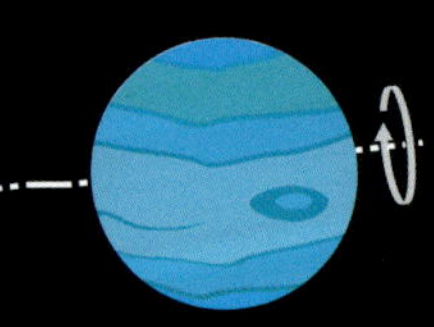

Year:
30,687 Earth Days
Day: 17.2 Hours

Neptune

Year:
60,190 Earth Days
Day: 16 Hours

The time it takes a planet to complete an orbit around the Sun equals one year. Planets closer to the Sun have shorter years than those far away. Planets also spin on their **axis**. One complete spin on an axis is equal to one day.

An atmosphere surrounds each planet. It is a layer of gases. Atmospheres can be thick or thin.

Thick atmospheres trap heat from the Sun, making the planet hotter. Planets with thin atmospheres are cooler.

Chapter 3

Rocky Planets

Rocky planets are closest to the Sun. Mercury, Venus, Earth, and Mars have hard, rocky **crusts**. Mercury is the smallest planet in the **solar system**. Venus is the brightest. It is also the hottest.

Earth

Mercury

Venus

Explore links here!

Mars

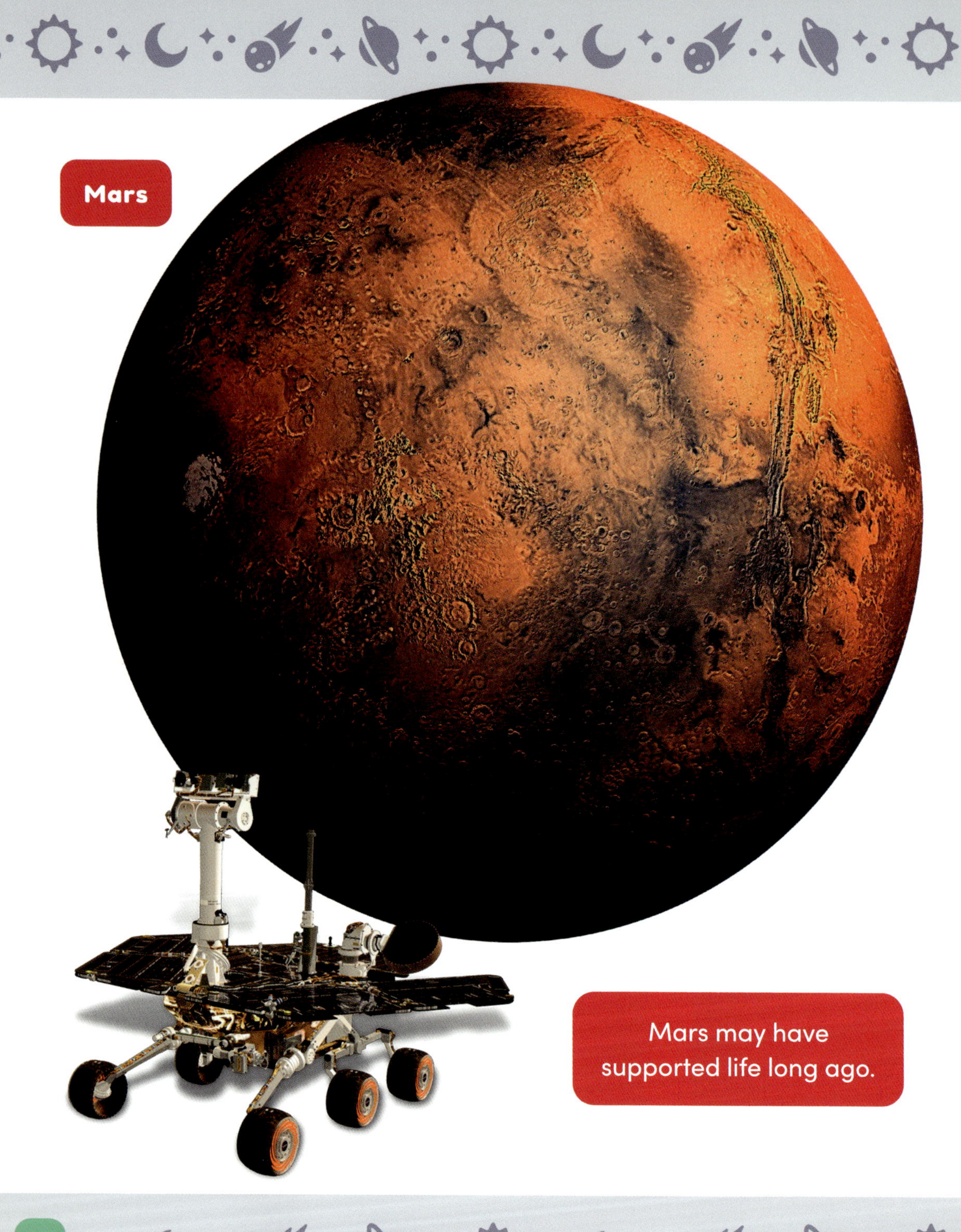

Mars may have supported life long ago.

Earth is the third planet from the Sun. It's the only planet that supports life. You can walk outside and see Earth's rocky ground. Mars is a cold, dry, and dusty planet. Scientists have sent machines to explore Mars.

Chapter 4

Gas Planets

Jupiter, Saturn, Uranus, and Neptune are called gas giants. These planets could have a small, hard, rocky center. But they are made mostly of gases, liquid, and dust.

Jupiter, Saturn, Uranus, and Neptune have rings.

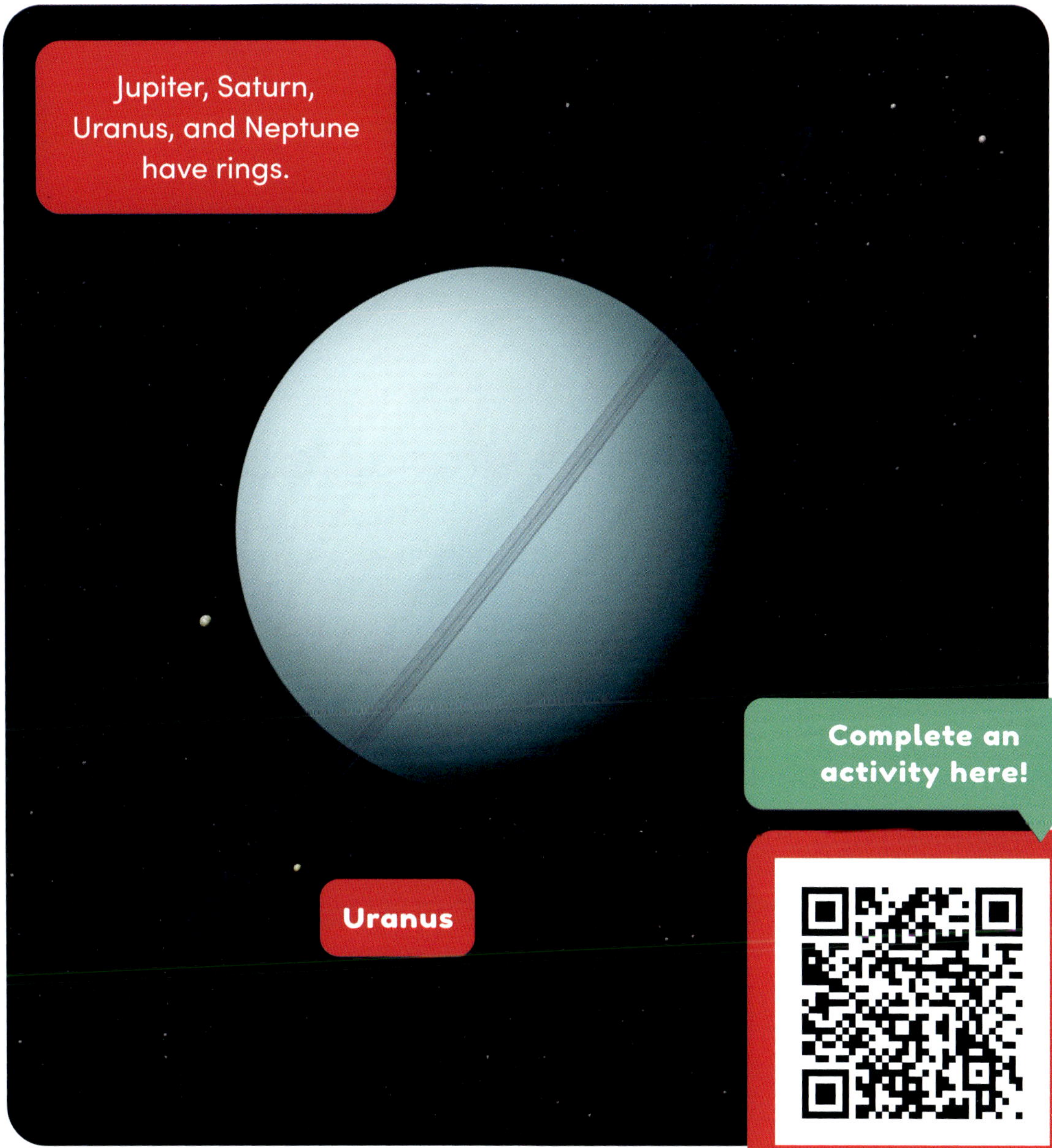

Uranus

Complete an activity here!

Jupiter

Jupiter is the largest planet in our **solar system**. It has a giant red spot on it. The spot is a 300-year-old storm!

Saturn is famous for its rings. The rings are made from chunks of rocks and ice. Uranus and Neptune are icy, gas giants. They are the farthest from the Sun.

Saturn

Making Connections

Text-to-Self

Besides Earth, which planet are you most interested in? Please explain your answer.

Text-to-Text

Have you read any other books about space objects? If so, how were those objects similar to or different from planets?

Text-to-World

Scientists study and explore space. Some are trying to find a way to send humans to Mars. Why do you think scientists want to visit Mars? Do you think there are other planets scientists should visit?

Glossary

axis – an imaginary line through the center of an object, around which the object turns.

crust – the hard, rocky outer layer of a planet.

gravity – a force that pulls objects toward each other.

orbit – the path of a space object as it moves around another space object. To orbit is to follow this path.

spherical – round in three dimensions.

solar system – any system that includes a star and all of the matter which orbits that star, including planets and moons.

Index

Online Resources

popbooksonline.com

Thanks for reading this Cody Koala book!

This book is filled with videos, puzzles, games, and more! Scan the QR codes* while you read, or visit the website below to make this book pop.

popbooksonline.com/obj-planet

*Scanning QR codes requires a web-enabled smart device with a QR code reader app and a camera.